Douglas Waterhouse

A GREAT AUSTRALIAN
...some memories of shared projects...

SEMISI PONE
BSc, MSc (Hons) Auckland

Publisher: Rainbow Enterprises Books 2020

ISBN: 978-1-98-851184-9

A short note on Doug Waterhouse.

Doug Waterhouse died on December 1, 2000 in Australia.
This short book is a tribute to Doug's efforts in the Pacific
Islands which we should never forget. Although, I cannot
recall much of our interactions, I do remember enough
about Doug especially his kind offer to help me write the
report for the Second Biocontrol Workshop at the
Mocambo Hotel, Nadi, Fiji in October 1995. That is
kindness I shall never forget. It does speak volumes
about the kind of man he was. I have checked the SPC
Library and Doug's report is there with Dr Paul Ferrar as
editors of the meeting papers. The report has been
available since 1997.

CONTENT

Chapters   **Pages**

INTRODUCTION

Dr Douglas Frew Waterhouse is one of the older gentlemen of Science in the Pacific Islands. Certainly, that was my view in October 1985 when I first met him. I felt that these influential older generation is a key ingredient in the success of any Science programme, in the Pacific, or indeed any programme. Doug had always given the meetings that he attended and projects he was involved in much kudos as they say in the Pacific, as well as a lot of credibility. I did not realize that when Doug came to Tonga in 1985, he was already retired for four years!

In the three meetings where I met Doug, over a period of 10 years, it was obvious that he was the centre of attention and source of the stories and discussions most of the time. I did not even think Biological Control was anything other than textbook material. Doug's stories had certainly made Biocontrol a kind of superstar career, in the Pacific Islands, with lots of fun chasing butterflies and moths with a fishing net in one's backyard.

Before the meetings in 1985 and 1995, biocontrol was not a topic people were aware of in the Pacific Islands. A lot of the media coverage in the local newspapers and radio popularized the often 'dirty' projects like rhinoceros beetle biocontrol which were usually associated with rubbish heaps in the main towns. Suddenly biocontrol of the rhinoceros beetle, in the rubbish heaps of Tonga and around the Pacific Islands, was a scientific quest equal to Ernest Rutherford's discovery of the atom.

Chapter One

The 1985 Biocontrol Workshop in Tonga
..my first meeting with Doug Waterhouse...

This workshop was held on 17-26 October 1985 at the Ministry of Agriculture, Fisheries and Forests Research Station at Vaini. There were two aims;

1. the background of biological control and how it should be practiced

2. Assess the suitability of the worst 23 pests and worst 15 weeds in the Pacific Islands which participants and country delegates will select for biocontrol programmes in their countries.

Dr Waterhouse went through these dossiers and although he had been retired since 1981 I can tell that he is still very active.
The workshop was organized by Dr Dirk H Stechmann and Konrad Engleberger of the Tonga-German Plant Protection

Project (GTZ) in co-operation with 'Ofa Fakalata and Vaea 'Anitoni of MAFF, Tonga.

Dr Paul Ferrar and Dr Doug Waterhouse also co-ordinated some of the action from ACIAR in Australia. A total of 56 country delegates, experts and observers from 20 countries attended.

I had just joined MAFF Tonga, five months earlier as an Agriculture Officer/Plant Pathologist so I attended many of the proceedings and presentations and especially the dossiers of pests and weeds presented by Dr Doug Waterhouse.

I was not one of the workshop participants as I was in the Vaini Research Station Plant Pathology laboratory and it was an Entomology Meeting. However, when ever I had time from my laboratory work I would check the meeting time table and attend Doug's discussion and presentation sessions. He was the star attraction at the meeting.

My aunty Lu'isa Kefu, my father's first cousin, took the notes for the publication of the proceedings and I did get a copy of it which I left in my Office in the Plant Pathology Laboratory in Tonga. I can tell that Lu'isa is very impressed with the meeting when I met her. She normally does the meeting records for big Government Meetings in Tonga being based in the Prime Minister's Office.

I must say that the social part of the meeting was a very joyous occasion especially the coffee time. You can hear everybody talking happily, laughing and catching up on the news when they have breaks from the meeting. One man stood out whose jolly personality endeared him to everyone, Mr 'Ofa Fakalata BSc, MSc, Government Entomologist. 'Ofa's belly laughter is so loud and infectious you cannot help laughing along with him. Certainly Doug thinks so. Whenever, I can hear 'Ofa from my office, I know its coffee time.

I met with some participants from the Pacific Islands who wanted a tour of town. I took them around town and met up with many of the others in the night life. There are some hilarious stories about the visitors that still remain to be told. That is how these meetings unfold in the islands. Pacific Islanders can work hard during work hours then have a few noisy parties, drinking beer or kava, after hours, singing and dancing to old traditional songs. It was the same all over the islands that I visited as part of my work for SPC.

Doug no doubt enjoyed meeting so many new people immensely, especially the Scientists from the islands. Dr Dirk Stechmann the main organizer of the workshop was very complimentary of Doug's huge contribution to the workshop. He normally picks me up from town on his way to the farm in the morning and likes to discuss those events during our half hour drive every morning.

From memory, the meeting reception was held at Kahana Beach Resort. I heard that some of the participants as usual, in the Pacific Islands, ended up at a Beach Party where one prominent member of MAFF was found sleeping on the roof of a faletonga (Tongan traditional house) at Ha'atafu Beach. That was the gossip at the Research Station for that week. It was all good fun and a bit of cheeky exaggeration.

I did not go to the reception because I did not get an invite, although the whole Ministry staff had attended with no invites. I guess my five years in New Zealand has changed me a bit. I did not want to be seen as a 'gate crasher' at the party!

I still remember some of the great results of Biocontrol projects in Tonga. Here they are;

1. Leucaena psyllid (*Heteropsylla cubana*)

This is one of Dirk's favorite subjects on our drives to the farm. The project started after I arrived in 1985. There had been a lot of MAFF planning for the Tonga-German Plant Protection Project as well as the workshop. It was decided to introduce the psyllid to Tongatapu to control the dense stands of Leucaena.

What was most noticeable was the rapid disappearance of the target species Leucaena (*Leucaena leucocephala*) after about a year. In two years the whole island of Tongatapu was completely cleaned of Leucaena. Only the dead stumps can be seen. It speaks volumes for the effectiveness of the psyllid and biocontrol methods. It is one reason why it is favored over the use of chemicals, in agriculture food production in some cases.

The psyllid biocontrol programme started off with complaints from some farmers that Leucaena was a problem weed in their gardens. Once Leucaena is established, they said, it is very hard to remove as the stumps are difficult to dig out and tractors cannot be used.

However, once Leucaena was completely destroyed, farmers began complaining that there is no firewood, fence posts, building materials and many other uses Leucaena was useful for in the villages in the east of Tongatapu.

The psyllid lays its eggs in the young shoots of Leucaena where the larvae eats and destroys the leaves. New shoots are destroyed quickly as they emerge. The Leucaena plants defoliate and die.

It was decided that since farmers can see a use for the Leucaena, they should be allowed to regenerate. In text book fashion, the psyllid population crashed when there were no more Leucaena to feed on. Our Scientists decided to allow Leucaena seeds in the ground to regenerate and never introduced the psyllid again. Dirk was always of the opinion that farmer's needs should be met and if they want the Leucaena back then they will not bring in any more psyllids for release.

In 2 years, dense stands of Leucaena can be seen all over the island of Tongatapu and farmers were grateful they were not completely destroyed. They now realise that Leucaena has more uses to them than its perceived problem as a weed.

2. Lantana (*Lantana camara*)

In the 1970s, the weed Lantana (*Lantana camara*) was a major problem in Tonga. The common story suggest it was introduced as an ornamental and has escaped into the wild and become a noxious weed. One often quoted negative feature of the Lantana was its 'prickly branches' which kids did not like as they get scratched trying to enter the bush to collect guava fruits or mango or even when clearing bush for the annual ma'ala (yam plantation).

Introduction of several bioagents into Tonga has had a noticeable effect on the Lantana. Usually guava bush stands around villages on Tongatapu is dotted with large clumps of Lantana as well as village fringes and cow paddocks. By 1991, very few Lantana plants can be seen around the island of Tongatapu. The ones that can be found often have the bioagents on them in great numbers, feeding on the leaves and killing the Lantana off before they get a change to get established in new areas.

3. Whiteflies (*Aleurodicus dispersus*)

Encarsia? haitiensis was introduced to control whiteflies in Tonga and around the Pacific with great success. The whitefly is one of the most irritating pests I have seen in people's gardens. It covers every plant in the garden making them look like they have been painted white, from a distance. Taro (*Colocasia esculenta*) and pele (*Abelmochus manihot*) plants, for example, can get infested by whiteflies and it would be hard the clean them off before cooking. A white powdery substance sticks to the leaves and does not come off easily.

Since the introduction of the biocontrol agent plants in our garden, for example, whiteflies quickly cleared up in a matter of weeks.

I noticed white flies on our hedge in New Zealand too, just last year, it was getting to the stage where the white 'powder' on the leaves stick to our clothes as we pass or touch the hedge. Great numbers of small whiteflies take off, when disturbed, creating a 'beehive effect'. Then all of a sudden they disappeared a few months later! It is the typical sign of predator effect in nature. There must be a biocontrol agent working in our neighborhood! Its the only explanation for such a sudden disappearance of a pest from an infested area.

The predators feed on the abundant prey or lay eggs on them and multiply to great numbers in a short period of time. As the predator number increases, the pest numbers decline in spectacular fashion.

4. Rhinoceros beetle (*Oryctes rhinoceros*)

Biocontrol of the rhinoceros beetle in Tonga since a decade earlier was considered a great success at the first workshop in 1985. Many of our MAFF staff at the Research Station were full time employees in the rhinoceros beetle eradication campaign all over the island of Tongatapu. One noticeable improvement was the absence of the tell tale 'fan leaf', on coconut tree crowns, in our coconut plantations. The beetle burrows into the base of the coconut crown and cut the new emerging leaves causing those 'fan leaf' because the cut leaves look like a fan. The ones women use to fan themselves in church on Sundays.

One just drives through the island and view the coconut plants by the road to check their health. In the past, before the rhinoceros beetle campaign, almost all the coconut trees by the road has some damage. After the virus was introduced to control the beetle, very few coconut plants by the road were damaged.

These are just some of the exciting stuff Doug and his mates do in Tonga and around the Pacific region. Although my work was involved with Plant Pathology projects I was fascinated by Dirk's stories of their biocontrol work.

Chapter 2

The Taro Beetle Biocontrol review
...Honiara, Solomon Islands...

As co-ordinator of the SPC Plant Protection Service I oversee the main issues concerning each project especially funding and staff problems. There were 7 major projects with budgets of more than $30 million. I covered those in the book PLANT PROTECTION IN THE PACIFIC (amazon.com).

I attend every review and also advocate for staff concerns such as housing and family needs and requirements. I make recommendations to management on staff housing needs. for example, if it is inadequate in some of the islands.

The review for the Taro Beetle Project was held at the main hotel in Honiara, Solomon Islands. It was the second review I had to attend after the Fruit Fly Project Review at the Warwick Hotel in Fiji's Hibiscus Coast.

We had one of the Meeting Rooms to ourselves. There were only the Project Staff , Dr Brian Thistleton, Dr Billy Theunis, Ioane Aloali'i, Roy Masamdu and myself and Doug Waterhouse, about 6 people in total. There were visitors, of course, from time to time.

The staff makes it really easy for me. They do all the work, I just agree with it and it goes to the donors and management to be approved. In this case it was a European Union (EU) funded project. I was the Manager of one of the EU projects with a budget of $NZ 5 million so I was well aware of EU procedures and processes.

Taro Beetle is a problem in most islands in Melanesia. So far, up to that time, it is not found in Polynesia. It causes cosmetic problems to export taro by boring large holes on the taro corm so they are unsuitable for sale although at the markets in Solomons and PNG, I have noticed some taro with some beetle damage. That is one point to be made about biocontrol, even if there is a little damage to target food species, it can still be used for food. Chemical control of pests can provide 100% localized eradication and good quality 'cosmetic appearance' of produce but very often chemical residue on vegetables, for example, were suspected to be higher than allowed by international treaties or conventions.

This is one area which was missing or neglected at the time. No one was testing and checking for chemical residue in Pacific Island produce, on a regular basis.

Doug's knowledge in Biocontrol was important to put our staff's achievements into perspective. It was crucial to have him as a 'sounding board' for our work. He has so many successful projects in the past and experiences that he can recommend certain bio-agents or procedures for the team to try out and it is also a learning experience every time to listen to Doug's stories.

Dr Brian Thistleton, the Team Leader of the Taro Beetle Project was an excellent Scientist and I requested his project presentation to take to the Marie Curie commemoration in Papeete, Tahiti, French Polynesia. I was invited to attend and just wanted to highlight the good work Dr Brian Thistleton and his team are doing and also create awareness in French and Eastern Polynesia on the problems of taro beetle, in Melanesia. Taro Beetle is not found in that part of the Pacific. As regional representative of the PPPO to the IPPC and FAO/UN consultations I am well aware of how we should handle regional pest movement.

No doubt the excellent control of Taro Beetle in many parts of the Pacific Islands today owes that success to some of Doug Waterhouse's contributions.

Chapter 3

The 1995 Biocontrol Workshop in Fiji
...Nadi, Fiji...

In the first workshop in Tonga 10 years earlier, it was recognized that the South Pacific Commission was the only organization in a position to do any work in the region. The Plant Protection Officer of SPC was tasked with doing the 'clearing house' work for dissemination of biocontrol information. It was also recognized there should be a Biocontrol Officer attached to the SPC Plant Protection Service. These were recommendations from the first meeting.

Being the Plant Protection Officer at SPC (1993-1996) and the organizer of the Second Biological Control Workshop at the Mocambo Hotel in1995, I was able to recruit a Biocontrol Officer prior to the meeting, Mr Albert Peters of Samoa.

It is interesting to note that in June 1994, SPC Management wanted all of the meeting dates and announcements for the next year and it was an exercise in crystal ball gazing because we had to put our trust in the plans and hope for the best.

I submitted all the meeting dates for 1995 and 1996. The meeting announcement went out to the member countries, at least, 4 months before the meeting dates. Some meetings were even more than 12 months away!

The observer invitations I had send out myself to organizations and individuals who are involved or interested in Biological Control in the Pacific. One of our project experts, the SPC/German Biocontrol Officer, for example, Dr Carlos Klein Koch had returned to Chile to take up a teaching position at a University there. I persuaded Dr Paul Ferrar of ACIAR to fund Carlos to join us at the meeting with a report of the status of Biocontrol in South America. Carlos did a great presentation at our meeting in Nadi. Some of the work done in biocontrol in South America were completely new to us.

I had already sent an invitation to Dr Waterhouse and no doubt the Australian Government would have the people who attended the first meeting informed as well.

It did occur to me at the time that Doug Waterhouse is a key person in all the activities going on and it so happened that Doug had seen me running around doing the meeting co-ordination and other house keeping duties he stopped me on the footpath one morning and said;

'Semisi, I know how busy you are, I can help by writing the meeting report. Don't worry I can do it'.

And I had a big smile on my face;

'Thank you Doug', I remember saying.

I was very relieved because the Plant Protection Service was having problems with meeting reports as no one wants to do it. We had some small talk as we walked towards the meeting hall. I realized at that time how well I have come to know Doug Waterhouse over the years.

Our publications was an area of huge concern because we have a long list of reports and publications waiting to be done. Some of which date back a few years. None of the staff can do it and I certainly have no time to do it. I requested approval from SPC Management to recruit Dr Grahame Jackson, former SPC Plant Health Officer, to complete the publications. We have money in my SPC/EU project to pay for it. Grahame had done many of the SPC-PPS publications during his time at SPC, which was a few years before me, and has the necessary experience we needed. It was a relief when Grahame took over the publications.

One person we missed at the Second Biocontrol Meeting was Dr Peter Maddison of DSIR, New Zealand. I remember Peter very well from the funny speeches he gave over the lunches during the first meeting and also other encounters. Peter was one of the more well known Entomologists in the region. I think Peter might have retired by that time (1995).

The outbreak of Coconut Scale in Tuvalu

One morning I found a request on my 'in tray' from our Manager, Dr Malcolm Hazelman. He says that it had been sent from SG (the Secretary General, George Sokomanu) and it need to be actioned. I read it, it was a request from the Government of Tuvalu. There is an outbreak of the Coconut Scale on the island of Nanumanga and they need help.

It occurred to me that our Biocontrol Expert, Dr Carlos Klein Koch had just left to take up a Lecturer job at a University in Chile. The new Biocontrol Officer had not been selected and there is only me and the Biocontrol Technician to do something about it.

I thought about Doug Waterhouse and the Australians. They have heaps of experts over there, but I decided after discussions with our Technician that we should just go ahead with it. It would take weeks or months to get an expert from Australia, New Zealand or Europe. I am familiar with the biocontrol procedures myself and the Technician, Mr Humesh Kumar, is pretty keen. He has done many collection and releases of the biocontrol beetles in Fiji.

I instructed him to collect the beetles and cage them in the lab. Once we have a good number we will go through the 'cleaning and safeguard procedures' then I will send him to Tuvalu with a box full of beetles.

The Tuvaluans were very happy when the Technician arrived and released the beetles. A few months later, the reports from Tuvalu suggest the Coconut Scale was 'declining in numbers'. There were follow up visits from other experts later but the initial release was very successful.

This is the announcement that went out to SPC member countries when the Coconut Scale was reported to be a problem in Tuvalu.

Ag Alert

OUTBREAK OF COCONUT SCALE IN TUVALU
A state of emergency has been declared by the Government of Tuvalu following the outbreak of coconut scale, *Aspidiotus destructor* Signoret (Diaspididae : Homoptera) on Nanumanga Island, Tuvalu.

Relatively dry conditions have favoured the outbreak which began in Tokelau Village in March 1994 and is spreading rapidly towards the new settlement at Toga Village.

Breadfruit trees are most severely affected, but coconut, banana and frangipani are equally damaged. Local treatments have included removing affected parts from plants, and burning or dumping them in the lagoon. This does not appear to have controlled the pest.

The Island Council has banned the movement from Nanumanga of live plants, green and mature coconuts and food baskets made from green coconut leaves in an attempt to prevent the spread of the insect pest to neighbouring islands.

Biological control organisms have been recommended and efforts are being made to obtain the predatory coccinellid beetles, *Chilocorus nigrita* Fabricus and *Cryptognatha nodiceps* Marshall, which will be released to achieve permanent control. The use of these coccinellids in Fiji has been successful. Coconut scale is recorded as having entered the Pacific Islands region in 1892 (Yap). It has occurred in an epidemic form in American Samoa, Federated States of Micronesia, Fiji, French Polynesia,

Guam, Marshall Islands, New Caledonia, Northern Mariana Islands, Paula, Papua New Guinea, Solomon Islands, Vanuatu, Wallis and Futurity, and Western Samoa The coconut scale has not been reported from Cook Islands, Kiribati, Nile, Tokelau and Tonga.
Source: Sandhu, G.S. (Sept 1994). Report of the consultancy of scale outbreak in Tuvalu, 1994: a report to the South Pacific Commission. 8 p. No. 12 26 October 1994 ISSN 1019-6226

Other biocontrol projects and IPM

There were many biocontrol programmes and Integrated Pest Management (IPM) strategies used by our SPC/German Biocontrol Project. Target species included 1. *Lantana camara* 2. *Mimosa in visa* 3.Diamond Back (*Plutella xylostella*) moth during my time at SPC from 1993-1996.

There were many outstanding stories from our 1995 Biocontrol Meeting mainly;

1. **Biocontrol Officer** - the first time a Pacific Islander was selected to be the Biocontrol Officer at SPC. Mr Albert

Peters is very experienced from Samoa, having worked on many projects there. Our panel thought Albert had all the credentials required for our project. Albert helped me run the 1995 meeting and also answer the questions from country delegates.

2. **Plant Protection Officer** - it was also the first time the Plant Protection Officer was a Pacific Islander (me), normally they were British, probably a tradition as most of the work done in the SPC-PPS since the establishment of the SPC in 1947 were done by the British. In 1947 there were only 6 member countries (Australia, France, the Netherlands, New Zealand, the United Kingdom and the United States).

I had a great time running the Plant Protection Service and as Doug pointed out, I was always very busy with 7 major projects and more than $30 million in budgets. My title was changed half way through my three years into Plant Protection Advisor. I was also the Co-ordinator for the SP PPS.

3. **Manager** - It was also the first time
the Agriculture Programme had a
Manager. Usually there were just 4
Officers (1. Plant Production 2. Plant
Protection 3. Animal Health 4.
Information) dealing directly with the
Director of Programmes (DP) in Noumea,
New Caledonia. Dr Malcolm Hazelman
was both Manager and Plant Production
Officer. Mr Poloma Komiti of Samoa was
DP at the time.

What was noticeable from the 1995
meeting was the younger generation of
the group. Most of the delegates were
young and so Doug had a special place as
the 'old wise one' when dealing with
question and answer sessions. The ones
who attended the 1985 meeting in Tonga
were older, on average. It was obvious to
us that Doug's presence had a huge
influence on the conduct of the meeting,
the science was first class and all the
participants were in awe of the man. It is
the main reason why I tried to get him to
come but Doug needed no invitation, he
would have come on his own accord

anyway. It was also the reason why I asked Dr Paul Ferrar to fund Dr Carlos Klein Koch, our previous expert with the SPC/German project, to come all the way from Chile. The presence of these older and more experienced gentlemen would have a huge influence on the young country participants and the meeting's success.

Chapter 4

Taro Leaf Blight Project
..the University of Technology, Queensland...

Taro Leaf Blight (TLB) was the biggest problem I faced as the new Plant Protection Officer for SPC. I took office in June and TLB was reported from the Samoas in July. A request came from the Governments of both Samoas to the SPC Secretary General which was passed to me for action. I looked up the disease outbreaks in other parts of the Pacific like Micronesia and Melanesia and it seemed the disease is no longer a problem in those countries. TLB had been introduced as long ago as World War II, into parts of Melanesia, and it has taken about 50-60 years to reach Polynesia. TLB resistant taro varieties already exist in Micronesia and Melanesia but quarantine problems like the presence of alomae and bobone viruses in parts of Melanesia will be a huge constraint to any future exchange of TLB resistant germplasm.

I thought the best approach would be to convene a meeting of the experts to discuss the problem of TLB in Samoa and American Samoa and propose solutions.

I booked a flight to both Samoas to discuss the problem and view the damage. I also took some pictures of the disease on the taro. I met the Governor of American Samoa and also the Director of Agriculture in Pagopago. I also met the Director of Agriculture in Apia and various high officials, I proposed to them that bringing all the experts together to view the problem and propose solutions might be the best way forward.

We convened the biggest meeting on TLB thus far, in the Pacific, at the IRETA meeting house, University of the South Pacific, Alafua Campus, Apia, Samoa. More than 70 country delegates, experts and observers attended the meeting. It was obvious that new information and ideas flowing at the meeting was very beneficial for staff of the Agriculture Departments from both Samoas. We also learned of the breeding work going on in the Solomon Islands and Papua New Guinea. I also got some ideas of how we can beat the TLB as soon as we can. Samoa was exporting $ST10 million before TLB of taro but now, a few months later, there is nothing to be exported.

I visited the Taro Breeding Programme of the Department of Agriculture in Lae Papua New Guinea. It was being carried out by Dr Anton Ivancic, Sim Sar and his staff. They showed me some of the successes they had and we also did some tasting sessions. It was all very edible from my point of view having had taro as part of my diet since childhood.

We discussed the possibilities of introducing the resistant cultivar to Samoa using a Tissue Culture Lab where they can be tested against alomae and bobone as well as Dasheen Mosaic Virus before introduction to the Samoas.

The Coffee Institute in PNG has a big Tissue Culture Lab and I went to visit them as well as their coffee research plots.

On the way back I visited the University of Technology in Brisbane, Australia to discuss the possibility of mass Tissue Culturing and testing TLB resistant taro from Solomons and PNG before introducing them to the Samoas.

Professor James Dale, Dr Rob Harding and Dr Brendan Rodoni were very supportive of my idea. They even offered to put together the project for ACIAR to fund. I was very grateful for their help. Like Doug they knew how much work I had to do. I also visited Dr Paul Ferrar at ACIAR in Canberra and discussed the possibility of introducing TLB resistant taro from Solomons and PNG to the Samoas through QUT funded by ACIAR.

I had worked with Dr Ferrar and his team on Kava Dieback in Tonga some years before so Paul was very happy to help. I told him I already have a team at QUT that can do the work.

Once I was back in Fiji, I decided to organise another big meeting for all the participants of the First Taro Seminar at IRETA. This time we will meet in Lae at the PNG University of Technology. Participants will have a chance to view the success of Anton and Sim Sar with Agriculture Staff in PNG. They have a number of very promising TLB resistant varieties. I was keen to introduce them to the Samoas as soon as we can get the funding and politicians on side.

I must point out that the alomae and bobone viruses reported by SPC a few years before was a huge stumbling block. The politicians cannot be persuaded that we can test for these viruses. In fact, it would be a good chance to develop a test for alomae and bobone at QUT. The team there were really keen to go ahead.
I did receive word from Professor James Dale, Dr Rob Harding and Dr Brendan Rodoni that ACIAR has approved $AUD

1 million for the TC Project. It was just a matter for SPC to bring the politicians from Solomons, PNG and Samoas together to agree to the exchange of germplasm and also for quarantine measures to be put in place such as the TC Lab and testing at QUT before introduction of TLB resistant TC taro to the Samoas.

The TC plantlets will be put under observation for a few weeks in green houses in the Samoas before planting material are given to farmers in both Samoas. I was very keen and I estimated that Samoa will start export taro again by 1997, four years after the epidemic in 1993.

The only problem was, it was clear by April 1996 that there has been no decision on my contract extension with SPC. Management had advertised all the core staff positions, as a 'management exercise' and no one seem to be in a position to tell me what is going on. I was on a 'end of contract leave' in Auckland

and I rang the new Secretary General, Bob Dunne to ask about my contract. Bob had just taken over from George Sokomanu and he says he has no idea what is going on but he promised to find out. Bob never got back to me on it and I guess no one knew what had happened! To make a long story short, I decided to stay in Auckland because I also had some health issues to sort out. I send Bob Dunne a fax saying I won't be coming back but I did not tell him about the health problem. Only my Secretary knew and I had instructed her to tell Malcolm.

I had called Dr Paul Ferrar at ACIAR and explained to him the problem with my contract. I recommended that it might be best they deal directly with the Government of Samoa. I may not be in Suva to co-ordinate the project.

As it turned out, it took 20 years before Samoa began exporting taro to Auckland again. The New Zealand Pacific Community was Samoa's biggest taro export market. The politicians did not agree to the exchange of germplasm as I had proposed. Although there were more than 200 accessions in the IRETA and SPC TC Lab collections, none were found to be resistant to the strain of TLB in the Samoas.

It did occur to me from the expert comments at the meetings in Apia and Lae that we can take the best resistant cultivars and test them in the Samoas after being cleaned at QUT. The TLB resistant cultivars were numerous. I could safely say, more than 20 in both Solomons and PNG. There were also some resistant cultivars in Micronesia. It is possible that , after testing for alomae and bobone, DMV and other microbes, we can mass produce each resistant cultivar into thousands and give a 100 to each grower in the Samoas (Samoa and American Samoa) to try.

It is possible for TC taro to be planted straight to the field after a few weeks in pots in the green house. We could train the farmers how to pot the TC taro then plant them into their fields. From the field plantings we can quickly work out the best cultivars under TLB disease pressure in the field. Within a year, it is possible for each farmer to produce thousands of planting materials themselves and select the best one themselves for export and local sales to earn income.

This would be a first from the point of view that farmers themselves will select the best plants for their farms. It has been the traditional way in the islands.

Taro is one crop plant that produce a large number of suckers or side shoots from the corm. These new suckers are used for new planting material. Growers can easily accumulate a large number of suckers, from just 100 plants, within one year.

I have seen the exported taro from Samoa in Auckland and bought them for our home cooking. They do taste great and I have seen the unloaded containers. I must say the bag design is brilliant. Anyone would think they are potatoes until they open the bags. I have also read the breeding work that went on in Samoa in more than 20 years and I guess it does not really matter whether they exported in 1997 or 2016. It's probably a good training experience for the local Samoan Scientists.

Chapter 5

The Kava Dieback Project
...a tribute to Dr Paul Ferrar...

Dr Paul Ferrar, Co-ordinator for ACIAR, was like Doug, always trying to help with our various Plant Protection problems in the Pacific. In fact, Paul and Doug ended up writing the report of the Second Biocontrol Meeting in Nadi, already mentioned, which is now in the SPC Library since 1997.

I had met Paul on many occasions but I wish to discuss our combined effort to solve the Kava Dieback problem in Tonga.

I was the Plant Pathologist/Virologist at the MAFF Research Station in Vaini, Tonga when ACIAR launched a project to solve the Kava Dieback problem which has plagued Kava Growers in Tonga for decades.

ACIAR put a PhD student, Richard Davis , in our Laboratory at the Vaini

Research Station to work with us. His
main mission was to solve the problem of
Kava Dieback for a PhD Thesis. I was
very busy with the Vanilla virus and later
Pumpkin virus. After about a year,
Richard asked me if I can have a look at a
Kava Plot near the farm. It was showing
symptoms but he has exhausted his ideas
testing mycoplasma, fungi and bacteria
but there is no result.

The symptom on the young Kava leaves
were chlorotic streaks and crinkling of the
lamina and necrosis on the stalk. I was
pretty sure it is a virus.

When Richard's project was reviewed, I
recommended to Paul, who was in the
review team with Professor John Brown,
that they should take some symptomatic
leaves and check the sap for virus
particles.

A week or so later, Richard excitedly
broke the news to me that the Kava leaf
sap was examined under the electron
microscope at the Australian National

University (ANU) and they found virus particles very similar to CMV.

We were all cerebrating. I ordered some ELISA test kits from Sigma USA with CMV antisera and we did a survey of Tongatapu, Ha'apai and Vava'u Kava plantations. Professor John Brown, Richard's supervisor, also came along with us to do the survey. Everything went very well in Ha'apai until one night while having a few beers in Vava'u, at the motel, John asked me how much I expect to be paid. I am obviously doing them a huge favor. I said, 'Well John, if you are buying all the drinks every time you come from Australia that will be good enough for me'. We have been having a few beers with John every time he comes to check on Richard's progress. I must say, I have never seen John laugh so much, so the beer kept on flowing that night.

The survey was a great success, all the 100+ samples with symptoms collected from Tongatapu, Ha'apai and Vava'u were tested positive to CMV using ELISA. It was the first time Richard had some success. CMV was definitely associated with the symptoms on Kava attributed to Kava Dieback, not only the virus particle was found in the symptomatic Kava leaves by electron microscopy, the symptomatic leaves also tested positive to CMV antisera by ELISA.

Obviously Paul was very pleased with the results. Richard went on to prove that CMV does cause the symptoms and kill the Kava plants through applying Koch's postulates and other techniques in his study.

Professor John Brown and Richard also did several surveys around the Kava growing countries of the Pacific, including Samoa and Fiji, funded by Dr Paul Ferrar and ACIAR. They also found CMV to be present in both Samoa and Fiji. They have published a number of scientific papers on Kava Dieback in the Pacific, one of which I am also co-author.

Paul was also very happy to meet me again at the Second Biocontrol Workshop. He was very, very supportive of Doug's vision for the widespread use of biocontrol in the Pacific Islands instead of chemicals. I guess there is a lesson to be learned there, that the islands would remain clean if less agricultural chemicals are used in the fight against agricultural pests.

I met Paul again when I visited to discuss the Taro Leaf Blight project at his ACIAR Office in Canberra. It would have been amazing to see the virus tested TLB resistant TC taro introduced to the Samoas and for export to commence just 4 years from the epidemic of 1993. I was very confident it will work and the added advantage of establishing a Pacific wide testing facility for alomae and bobone viruses as well as dasheen mosaic virus (DMV) which is a low priority and widespread disease of taro and cocoyam in the Pacific.

The QUT will also host the TLB resistant elite virus tested taro collection *in vitro* at its facility for the Pacific Islands. The collection will also be used to supply Pacific Islands that need TC taro for hurricane recovery and research projects.

Although we had facilities for TC and popular *in vitro* collections of taro, sweet potato, yam, banana at USP and SPC, at the time, I felt that a QUT Lab to support those works would be perfect, especially in diagnostics which was pretty much absent in the region.

It wasn't to be and I guess that is the way it is and remains to this day.

Chapter 6

The RTMPP9 and PACINET

..a tribute to learned colleagues...

Being the Regional Director of the South Pacific Commission Plant Protection Activities (1993-1996) was a huge responsibility with a heavy workload.

It was fortunate that on many occasions many learned colleagues from across the region offered to help. I wish to mention some of them in this short book to honour our best Scientists in the Pacific, certainly during my time there.

Our three meetings at the Tanoa Hotel Fiji, during February 1996, is the culmination of my 3 years as SPC Plant Protection Advisor which was a role with many titles.

They include the following;

1.   Co-ordinator of the SPC Plant Protection Service with 7 major projects, 24 staff and more than $NZ30 million in budgets.

2. Head of the SPC Plant Protection Service

3. Chief Executive of the Pacific Plant Protection Organization

4. Manager of the SPC/EU Pacific Plant Protection Project with $NZ5 million in budgets.

9th Regional Technical Meeting in Plant Protection

The RTMPP is the oldest meeting in Plant Protection in the Pacific region having run continuously, every 2 years, for the previous 16 years or so. My role was to send out the meeting announcement and agenda a few months before the event announcing the venue and date. I write up the 'Savingram' and send it to our Editing and Interpretation unit in Noumea, New Caledonia, who produce 2 edited copies in English and French, approved by Management which include the Secretary General (George Sokomanu of Vanuatu), Director of Programmes (Poloma Komiti of Samoa) and Director of Services (Fusi Taginavanua of Fiji).

The approved Savingram is then sent out to member countries in English and French.

Normally at the meetings, my staff will help me to run the meetings and also write the reports. Unfortunately, I was unable to hire somebody with enough experience to write the meeting reports, on time. Although I had $NZ 5 million which include staff salaries for a Biocontrol Officer, Information Officer and Librarian, they only came on board in 1995. The staff recruitment process takes months or even years to complete.

For those who are not familiar with writing SPC meeting reports, there is a lot of scientific information and details to gather before you can write a meaningful report.

It so happens that at the RTMPP 9, we did not have any staff in the Plant Protection Service Secretariat, experienced enough, to do the reports although it was part of the job description of some of our new staff, who have only come on board for a few months.

To be fair, I had to do all the reports myself by recruiting experienced country delegates. In the the case of RTMPP 9, I found some kindred spirit in the New Zealand Delegation. Richard Ivess and John Hedley had been strong supporters of the SPC Plant Protection Service over the years and they were keen to help me.

We got some coffee and sat in Richard's room and type up the report til late that night. It will be presented to the 27 member country delegates to the SPC RTMPP 9, the next morning. I must say, Richard was the right choice as he had the latest laptop and printer money can buy! I did not even have a laptop myself! Richard was the Chief Plants Officer in the New Zealand equivalent of a Ministry of Agriculture in the islands so he has the connections! Laptops were very high tech and expensive in 1996 being very new to the market. I bought mine, which was a Mac, later that year, in Auckland, for $NZ 5,000! Compare that with the one I am using now which is a HP valued at $NZ 600, twenty four years later!

Anyway, Richard, John and the New Zealand Delegation were my saviors that time. We had a pretty good report written up, printed, photocopied by morning. All 70 delegates, observers and visitors got a copy in their pigeon holes by the time the meeting convened at 9am.

PACINET (Pacific Biosystematics Network)

The PACINET page on the SPC (Secretariat for the Pacific Community) website says;

PACINET (the Pacific Islands Network for Taxonomy) is designed to build taxonomic capacity in the Pacific Island Countries for sustainable development. It is a joint program of the Secretariat for the Pacific Community, the Secretariat of the Pacific Regional Environment Programme and the University of the South Pacific. Taxonomic capacity is having the human resources to identify, describe, name and classify the unique biodiversity found in all the Pacific Islands landscapes including rainforests, grassland, freshwater systems, atolls, coral reefs and the deep ocean.

PACINET's activities are based on the needs of PICs, and its annual work plan and business plan are overseen by a steering committee made up of a Locally Organized and Operated Partnership (LOOP). PACINET has a fulltime Coordinator who oversees the day-to-day operations and implements the work plan.

PACINET works closely with the BioNET-International (www.bionet-intl.org) - a global network to promote taxonomy in the biodiversity rich but economically poorer countries of the world.

Overall, PACINET will focus on providing coordinated access to existing taxonomic information and increasing taxonomic capacity in the region. PACINET will endeavour to facilitate and strengthen links between modern (scientific) taxonomy and local (vernacular or traditional) taxonomy as a foundation for improving the conservation, sustainable use and equitable sharing of the benefits of biodiversity in the Pacific region.

PACINET welcomes contributions from individuals, private sector, government and non-government organizations in this quest to build regional taxonomic capacity to secure the future of Pacific Islanders.

Contact person: Dr Mary Taylor

Dr Mary Taylor was my colleague at the University of the South Pacific in Samoa.We were responsible for the Tissue Culture laboratory. I considered Mary to be the 'Senior partner' as she has been there before me and had worked so hard to establish the laboratory. It was a 'world class' facility and I would credit Mary with 'most of the work' in its establishment. I had published my work at the USP TC Facility in my book PLANT PROTECTION IN THE PACIFIC 3, Tissue Culture (amazon.com).

I think Mary is the right choice to look after the PACINET loop which collaborates with other loops around the world, as well as BioNET International.

I have already discussed the PACINET project in some detail in my book PLANT PROTECTION IN THE PACIFIC. It was actually kicked off my my predecessor, Mr Robert Mcfarlane and I simply took over from him.

My first decision as the new Plant Protection Officer was whether to support PACINET or cancel it. I read through all the available information and I decided to support it. My motivation was the possibility of training and providing a taxonomic service for the islands of the Pacific Region. I travelled to the University of Cardiff, Wales, to participate in the Global Meeting and also to find out more about it. Professsor Tecwyn Jones and his co-ordinating team at CAB International were keen to meet me and find out what I think. PACINET was certainly an important part of the proposed BioNET International initiative.

The decision was made we will have to do all three meetings;

1. RTMPP 9
2. PACINET
3. PPPO

in February, 1996 at the Tanoa Hotel because my contract ends in May and I still have about one month of holiday which I need to take before the end of my contract. There will be no other time for convening the meetings.

Professor Tecwyn Jones, of BioNET International, who had invited me to the Global Meeting at Cardiff had promised $US100,000 to help finance the PACINET meeting. My staff will provide the Secretarial support.

It was the main reason why we had 3 large meetings in one month! All the reasons are right and people are keen! I was going on leave and I have run out of time.

My job was advertised by management as a 'restructuring exercise' and I was not sure whether I will have a job in 3 months time. All the core budget jobs were advertised. They were the jobs funded from country contributions.

It was a terrible feeling. I had worked so hard to get more than $NZ30 million for the Plant Protection Service projects, in just 3 years, and it looks like somebody else will come and spend the money for me. I was working in a foreign country with a wife and 2 young kids and it didn't look like the best place to be for my kid's future.

Normally we would have one large meeting every two years! Having 3 large meetings in one month and getting everything done smoothly was resounding proof of the calibre and hard work of our team. Everyone were very impressed by how smooth the 3 meetings had run during those two weeks.

Professor Jones was also very helpful with his staff. He had designated one of his staff to do the meeting report especially the one for the donor.

We had a planning exercise with all 54 Country Delegates and come up with a $US8.3 million budget for PACINET at the end of the meeting. Professor Jones and his staff put together the draft report which we all read and added to before it was sent to the donor, the Global Environment Facility to approve. That was another report that I got some help with.

I had realized in late 1993 when we had the large meeting on taro leaf blight in Samoa, that if I were to get anything done I should enlist the help of the country delegations. I have the choice of Australia, France, New Zealand, United States of America and the United Kingdom to call upon for help; in addition to 22 island states and territories.

There were also staff available through projects from Germany, Chile, United Nations and collaborators on the projects. My problem at the beginning was, I did not have my EU project staff to help me until 1995. For two years, since I started in June 1993, I managed on my own with about a third of the staff who were mostly technicians.

In order to make the PACINET project successful, I managed to get everyone on board before the meeting. We invited two delegations from each of the 27 member countries. One from the environment department and one from the agriculture department. That is why we had 54 delegates as well as observers and visitors at the PACINET meeting. It was the only meeting I organized with 2 delegations from each country.

We also invited regional stakeholders like SPREP, USP and many non-governmental groups to attend.

The $US 8.3 million budget was approved and funds made available after I had left SPC in 1996.

Abbreviations

1. ACIAR - Australian Centre for International Agriculture Research

2. ANU - Australian National University

3. CAB International - Commonwealth Agriculture Bureau International

4. CMV - Cucumber Mosaic Virus

5. DMV - Dasheen Mosaic Virus

6. DSIR - Department of Scientific and Industrial Research

7. ELISA - Enzyme Linked Immuno-Sorbent Assay

8. EU - European Union

9. FAO - Food and Agriculture Organization of the United Nations

10. IPPC - International Plant Protection Convention

11. IPM - Integrated Pest Management

12. IRETA - Institute for Research, Extension and Training in Agriculture

13. PACINET - Pacific Biosystematics Network

14. PNG - Papua New Guinea

15. PPPO - Pacific Plant Protection Organization

16. RPPO - Regional Plant Protection Organization

17. RTMPP - Regional Technical Meeting on Plant Protection

18. SPC - South Pacific Commission

19. SPC PPS - South Pacific Commission Plant Protection Service

20. TC - Tissue Culture

21. QUT - Queensland University of Technology

22. UN - United Nations

23. USP - University of the South Pacific

24. USA - United States of America

About the author

Semisi Pone was the Plant Protection Advisor and Co-ordinator for the South Pacific Commission's Plant Protection Service from 1993-1996. During that time he was 1. Manager for the SPC/EU $5 million Pacific Plant Protection Project 2. Chief Executive for the Pacific Plant Protection Organization 3. Member of the United Nations FAO panel of experts on Biosecurity 4. Member of the United Nations FAO Technical Consultation among RPPO's.

This is the second edition of this book which was written to include other important projects and scientific work in the Pacific. And also include a tribute to all those colleagues who made his work in the Pacific Islands (1985-1996) such as success.